Quantations

Quantations

A Guide to Quantum Living in the 21st Century

Joseph A. Stirt, M.D.

Writers Club Press

San Jose New York Lincoln Shanghai

Quantations
A Guide to Quantum Living in the 21st Century

Writers Club Press
an imprint of iUniverse, Inc.

For information address:
iUniverse, Inc.
5220 S. 16th St., Suite 200
Lincoln, NE 68512
www.iuniverse.com

Read only if you want to change the way you think about reality forever.

ISBN: 0-595-21159-3

Printed in the United States of America

Only the incomprehensible is worth understanding.

One must always act on insufficient evidence. Even the most important decisions in life must always contain an inevitable amount of irrationality.

All human experience is ordinary, not quantum. Whether our reliance on classical modes of perception is a permanent feature of the human condition remains to be seen.

We cannot hope to do better in the future by finding a deeper level of reality.

God created time out of eternity.

Death is simply the collapse of the superpositions we call "life." The ultimate decoherence.

Look at it in a quantum light.

The conscious mind is just catching up on past information.

Matter is thought, and thought is matter.

I could be bounded in a nutshell and count myself king of infinite space.

Everything is not relative; rather, it is all real and happening simultaneously.

All things will be in everything; nor is it possible for them to be apart, but all things have a portion of everything.

Nonsense and beauty have close connections.

We live between the inexplicable and the unpredictable.

Explanations are clumsy rationalizations with hindsight.

It is humanly impossible to come to terms with the full extent of the randomness, luck, and accident involved in life.

There are many coexisting truths, from which we must pick and choose.

We must decide between <u>and</u> and <u>either/or</u>. For the Greeks they're just two sides of one coin.

Conscious life is the experience of quantum reality.

All the different ways history might have unfolded exist in superposition.

Take a living and a dead thing, keep increasing the magnification, and eventually you won't be able to tell which is which.

In biological terms, life has no goal. We want nothing more than to move forward, glutted with goals, strategies, and challenges. But there is no movement forward. There's not even movement backward. There's only movement.

There is no such thing as chance, and what we regard as blind circumstance actually stems from the deepest source of all.

The future enters us and transforms us long before we are aware of its presence.

Time is downloaded into our bodies. We contain it. Not only time past and time future, but time without end. We think of ourselves as singular and finite, when we are multiple and infinite.

Man is always ready to ascribe human motives and intentions to phenomena at large, most notably in his almost obsessive search for meaning in the timing of events in everyday life.

Converts are not won to a religion or even a worldview by uncertainty and complexity.

We live in a through-the-looking-glass world in which yes and no can be true at the same time, and where events can happen simultaneously in multiple places.

In the quantum world, common sense, as developed in our large scale experiences, ceases to apply. Particles appear less as things of substance and more like waves, with the disarming potential to be in more than one place at a time.

Consciousness is a subtle form of matter.

The quantum self is the notion that what you are experiencing now is not the ultimate; that you are part of something that is much more interesting and complex than your own personal life.

Daily life is a 30 frame per second collapse of superpositions; alternatively, our consciousness is a light racing down tunnels, veering off 30 times a second.

The world is not objective but, rather, contingent, made up not of truths but rather of options and scenarios.

Things and facts and phenomena of daily life are real, yet the atom and its elementary particles themselves are not: they form a world of potentialities or possibilities rather than one of things or facts.

One reality does not displace the other but lives inexplicably alongside it.

What is it that governs our lives? Is is chance? Or is there some controlling presence behind the visible reality?

Quantum tunneling is the bizarre ability of particles to penetrate impenetrable barriers. It is the basis of modern electronics systems and computing, so it does occur.

30% of the U.S. gross national product is based on inventions made possible by quantum mechanics, from semiconductors in computer chips to lasers in CD players, MRI machines in hospitals, and much more. The quantum world, in all its weirdness, is real.

A barrier placed in the path of a tunneling particle does not slow it down. If you increase the thickness of the barrier, the tunneling speed increases. The tunneling speed apparently greatly exceeds the speed of light.

Humans can never experience the true texture of quantum reality because everything we touch turns to matter.

Only atoms and empty space have a real existence.

To use quantum memory, scan back to where you were when you thought of something: the something, being part of that particular past, will come back too.

Magnetic resonance imaging scrambles quantum states, yet patients emerge from MRIs seemingly unchanged.

If you can't measure a difference, there is no difference.

Affect is the algebraic sum of decohering feelings.

Traditional African thought tends to construe the unconscious as a force-field exterior to a person's immediate awareness. It is not so much a region of the mind as a region in space, the inscrutable realm of night and of the wilderness, filled with bush spirits, witches, sorcerers, and enemies.

Consciousness is a basic property of the universe, like space and time.

We tend to think of memory as a camera or tape recorder, where the past can be filed intact and called up at will. But memory is none of these things. Memory is a storyteller, imposing form on a raw mass of experience, creating shape and meaning by emphasizing some things and omitting others. It finds connections between events, suggesting cause and effect.

The ultimate impediment to a network's efficiency is the speed of light.

Quantum thinking eliminates both doubt and certainty. We believe nothing, yet realize everything must be so.

Coincidence is a glimpse of the scaffolding of reality.

Contagious magic is based upon the assumption that substances which were once joined together possess a continuing linkage; thus an act carried out upon a smaller unit will affect the larger unit even though they are physically separated.

Quantum reality consists of simultaneous possibilities.

Quantum physics is a kind of code that interconnects everything in the universe, including the physical basis of life itself.

To change the way we perceive the world as humans, we must become other than human.

The more intense the emotions, the lower the level of consciousness, and vice versa. A drug, a dance, a bungee jump temporarily obliterates the mind.

Chaos is the energy source of the universe.

Life is but thought.

I do not know if it has ever been noted before that one of the main characteristics of life is discreteness. Unless a film of flesh envelops us, we die. Man exists only insofar as he is separated from his surroundings. The cranium is a space traveler's helmet. Stay inside or you perish. Death is divestment, death is communion. It may be wonderful to mix with the landscape, but to do so is the end of the tender ego.

Consciousness is a small-scale black hole, drawing into itself intangible stuff and emitting not x-rays or quasi-stellar radiation but THOUGHT.

Quantum mechanics adopted an accepting posture that opposites could both be true at once. Suddenly the world made sense.

Our judgment does not seem to be the product of logical analysis; it wells up from somewhere beneath our consciousness.

An entity exhibiting quantum computing is conscious. Quantum decoherence yields consciousness.

Matter and consciousness meet in the quantum world: to attempt to distinguish them is silly. Let's just say that matter is conscious and move on.

The atom is not a hard permanent structure, but mostly empty space. Matter which is solid to the touch is transparent on an atomic scale.

Old science sought ultimate truths and fundamental laws; new science looks for unexpected possibilities and emerging patterns.

Interaction between individuals consists of decohering temporarily into the same quantum collapse.

In quantum reality there are millions of possible worlds, unactualized, potential, perhaps bearing in on us, but only reachable by wormholes we can never find. If we do find one, we don't come back.

Scripts, sets, personae, roles: all reflect the quantum indeterminacy of life today, which only collapses as necessary into a given reality.

Quantum man must live with ambiguity, not search for certainty. Competing, conflicting goals, priorities, and truths are par for the course; they are the stuff of life, the DNA of postmodern consciousness.

There is no particular justification for why things are the way they are. Any number of arbitrarily small perturbations along the way could have made the world as we know it turn out very differently. We are forced to admit that the world as we know it is the result of a long string of chance outcomes.

The law of causality is no longer applied in quantum theory.

Objects carry with them a load of received consciousness. Lightening the load of possessions and responsibilities frees attention.

The power of a great idea does not lie in its ability to solve an obvious problem, but rather in its ability to create a market based on latent or hidden needs.

If each neuron in our brain must be "on" or "off" in response to 10,000 to 20,000 impulses at any given instant, decoherence which results in large scale behavior can only be termed magic.

Emotion, the most basic form of consciousness, lies at one end of a continuum; the other end is occupied by the mind.

What might be fades in the world of necessity. Creation is gathering smoke. Reality fights against dreams, attempting to deny change and emergence.

We lead our life in an intellectual, apparently logical way but, in fact, our lives are irrational and there are things we can't explain. We are surrounded by a wealth of possibilities, and yet feel profoundly there is something missing.

Man has a nearly universal urge to believe that behind the otherwise inexplicable workings of probability is an all-knowing God who has a reason for everything that he chooses to happen to us.

Superpositions have a bizarre property: there is never only one of them. In fact, you could say they are "entangled."

What would quantum personality look like? For one, moods and emotions would bear no relation to events; you would feel good or bad for no apparent reason. Of course, you could associate certain things to feeling one way or another.

How much of mood and emotion is simply noise, randomly appearing, then ebbing away; where is our true nature? Is there an essential identity within each of us?

It's only because we live so sunk in ourselves that we don't notice that what is actually happening to us leaves intact, at every moment, what might happen to us.

In the unpredictability of the day, of the life, lies its promise: to know what is to come is to already have experienced it.

The ultimate quantum gesture is the shrug.

The quantum reptile is the turtle. The quantum bird is the crow. The quantum insect is the dragonfly, a rainbow in motion. The quantum flower: the daisy.

As long as you don't ask, you'll never get "no" for an answer.

In the realm of atoms and smaller particles, objects exist not so much as objects but as mists of possibilities of being here or there or everywhere at the same time: the quantum fog.

The probabilistic nature of quantum events means that if a stream of particles encounters an obstacle, a few, conveyed by probability alone, will magically appear on the other side of the barrier.

Everyday life is different from the quantum world. Objects--buses, tacks, friends--exist at one place at one time. Either they are here, or they are not.

The illusion of time is to make us think we are moving forward into the next act, life being an infinite series of acts.

We cannot see time, but we can look back through time at ancient images from distant objects in space. We live in a present surrounded by spherical shells of the past.

Life is a finite period of a finite series of acts, or scenes.

To explain the world completely, matter and energy are not enough; information is required. These are the three foundations of our world.

The principle of life cannot be destroyed. It is written into the cosmic code, the order of the universe.

Quantum intuition is what I call our sense of what we believe regardless of logic and fact.

Quantum reality can thus be viewed as ordering the universe, from the possible to the merely more likely.

Thought can exceed light speed. Imagine yourself on the moon right now: you see yourself there instantly, yet light takes 1.5 seconds to arrive. Thought moves infinitely fast.

Life is a near-death experience.

All work and all love, the search for wealth and fame, the search for truth, life itself, are made up of moments which pass and become nothing. Yet through this shaft of nothings we crive onwards with that miraculous vitality that creates our precarious habitations in the past and future.

One must have faith in illogicality, the confusion of rules and the impotence of intelligence.

Everything is true; the opposite is true too; you must believe both equally or be damned.

When you can't find something, does it disappear and then return when you find it?

Intellectuals have a compulsion to be right. This urge is inimical to what Keats called "negative capabil.ty," the capacity a poem or novel (or consciousness) must have to keep afloat a thousand questions in order to create the parallel universe of art. Rationality, to borrow from Foucault, may be an inferior level of discourse.

The body is one of the things in which our true feelings are located, but it is not the only one. Least of all is the self limited to the body. A person literally projects or throws himself out of the body, and any-where at all.

When someone invests psychic energy in an object--a thing, another person, or an idea--that object becomes "charged" with the energy of the agent.

Objects are more fundamental than time. The universe at any given instant simply consists of many different objects in many different positions.

Many entities that seem to exist are not part of the real world, like a rainbow or a mirage, or a mirror image.

Death has a hard time gaining a foothold in a universe where everything can be converted through magic or language into something else; nothing can ever really be lost.

You were dead billions of years before you were born.

Dividing the universe into living and nonliving things is meaningless. Even a rock is in some way alive. Life and intelligence are present in all matter, and in energy, space, and time.

Time present and time past are both perhaps present in time future, and time future contained in time past.

Human consciousness collapses one lucky universe into being from all of the possible ones.

The flow of time from Big Bang to life, death, and the eventual fate of the universe itself may be in a sense merely an illusion in the macroscopic world, whereas at a quantum level time appears to flow symmetrically in either direction.

Entropy happens in an instant; order takes time.

The transition from possible to actual takes place during the act of observation.

Nothing is nothing.

The truest and most evocative fictional characters take shape from the pinpoints of who they are, and from the undermining, even contrary flow of who they might otherwise be or possibly become. They are both particle and wave, simultaneous possibilities.

What exists and what might exist are windowed together at the core of reality. All the separations and divisions and blind alleys and impossibilities that seem so central to life are happening at its outer edges.

Religion, morality, and common sense will continue to deter people from truly believing in the ever tinier discoveries of science. More importantly, particles suggest mortality. People still fear dust's final requiem for life, still dread the infinite granularity of things.

Fundamentally, is the flow of time something real, or might our sense of time passing be just an illusion that hides the fact that what is real is only a vast collection of moments?

"Nothing" is the force that renovates the world.

Once the improbable has been ruled out, the solution must lie in the realm of the impossible.

That matter is composed not of inert, solid particles but of waves, fields, and probabilities means that matter is intelligence or spirit.

The "primordial soup" existed for 300,000 years after the Big Bang, until atoms formed. Before the "primordial soup" and the Big Bang, a vacuum existed, or nothing. For the universe to come into being, nothing is something. Nothing has energy. Nothing can change.

Certain moments have a gift of revealing the past and foretelling the future. It is these moments that I hope to catch. My wish is to make something permanent out of the transitory.

My wish is to make something transitory out of the permanent.

Physical objects of a shared living space are "solidification of void."

Apparently empty space is actually a sea of what scientists call "virtual particles" that briefly appear and disappear.

Time is the most important and the most enigmatic property of nature. Time is not propogated like light waves; it appears immediately everywhere. Time links us and all things to the universe. It carries information over any distance instantly.

What is it like to experience the world as an animal does? When we look around, things catch our attention as they move or take on familiar shapes. We recognize things for a brief instant before the word comes to the fore.

Man has invented time. Animals don't know what time is. We must accept that there is one big now, that the future and the present are not separate. The shadows of the future are seen in the present.

That which constitutes the logic of our normal acts and our normal life is a continuous rosary of recollections between things and ourselves and vice versa.

The universe itself can be described as a wave function, all possible histories hovering together in superposition.

The world can be built more or less of structured nothingness. Force and matter are manifestations of space and time.

The past and the future are just special cases of other universes.

Empty space is never completely empty, but instead burbles with "virtual particles" that wink into existence and then vanish before they can be detected. Though ephemeral, virtual particles still exert pressure by bouncing off objects during their brief lifetimes.

At the root of existence, matter is pervaded by mind. The unpredictability of electrons is mirrored by the sensation we call free will. The brain (and mind) have evolved to take advantage of this elementary freedom. You can call it freedom or you can call it a random process. But the word random is always used as a cover for ignorance.

Religion occurs when our imagination becomes powerful enough to extend belief to nonphysical entities.

Magic, religion, and science are nothing but theories of thought; as science has supplanted its predecessors, so it may hereafter be itself superseded by some more perfect hypothesis. The dreams of magic may one day be the waking realities of science.

Time is a shuttling of the future into the past, moving through an immeasurable point.

Time exists only for humans, as does mystery. Once the past is used to prepare for the future, time becomes necessary.

There is no past and no future. Time and motion are illusions. Every moment of every individual's life--birth, death, and everything in between--exists forever.

Self-awareness is a continual spanning of two points, the storable future and the stored past.

A human experiences himself, his thoughts and feelings, as something separated from the rest, a kind of optical delusion of his consciousness. This delusion is a kind of prison, restricting us to our personal desires and to affection for a few persons nearest us.

Without eyes you have less of a present, without a nose you have less of a past. The more past you have, the longer you live; a person is as old as his oldest memory.

Since at the speed of light time stops, from the point of view of a photon, the moment it was created 13 billion years ago and right now, when it is observed, and forever into the future, as it travels, are all the same instant.

The apparently accidental association of events in the external world we call coincidence. Actual human experience is unified by repetition of events, physical and mental. The mind relentlessly associates present experience with suggested fragments of the past.

The only truths and facts are those which stem from observation of human nature, which has never changed and never will; all else is passing fancy and modern theory, and is left in fragments alongside the path of time.

In collapsing space and time, linking distant events in both spheres, Proust was the first of the great quantum thinkers; he shows us entanglement and action at a distance in every moment of everyday life.

Since mass and energy are equivalent, we may see that all elementary particles consist of energy, which defines energy as the primary substance of the world. Energy is that which moves, and is the root cause of all change.

Space has always existed in the complete shape of the universe in four dimensions. It is only because of the way the human mind works that our consciousness insists on experiencing the universe one moment at a time [30 frames per second].

The Corinthians asked, "What are the things in life that never change?" Paul answered, "The things you cannot see."

You attribute time passing to events which appear to you in the instants; what belongs to the form you carry over to the content. You hardly see this amazing flow of time. You see a woman and at certain moments you think you see her grow old and feel yourself growing old with her.

The difference between reality and illusion, say the postmodernists, is that the more powerful experience is the real one. Intensity is all. It is an "aesthetic hallucination of reality."

In saving the appearances, quantum theory jettisons the apparent reality of everyday life. Cause and effect gives way to pure randomness, laws of logic are flouted, things--if things can be said to exist at all--possess no determinate properties until a conscious observer appears.

The Western objective view is that matter and energy, after billions of years, became conscious. The Eastern subjective view states that consciousness came first, and that matter and energy are merely the complex thoughts of conscious beings.

Man might not be much smarter, or fundamentally in more control of things, than a rock, since everything is made of the same particles and governed by the same uncontrollable forces.

Nothing completely disappears, everything is transformed; what we believe to be dead has but changed places. What is, is thought. What is thought, is. Everything contains the aura of what it previously was, and the aura of what it will be when it disappears. You belong simultaneously to the present, past, and future.

Consciousness may arise from quantum mechanics, the same process that governs the behavior of subatomic particles.

The universe as we see it is a paradigm of our lives-a very definite beginning with an indefinite, but inevitable, end.

New and old do not exist in the world, because everything in it is now.

The painter's task is to imagine the world as it was before it had been converted into a network of concepts and objects.

I do not have a past or a future, only a continuous present.

Magic/superstition: control forces of the universe
Religion: surrender to forces of the universe
Science: understand forces of the universe
Quantation: become forces of the universe

Time does not exist. Time is a convenience which lets us say things are different without contradiction. Without time we must put things in different worlds.

Language must make sense on an atomic scale, as well as in the classical world. Therefore mathematics is not the "lingua franca" of the 21st century. Rather, language must evolve into a supple quantum medium.

Fear of the incomprehensible impoverishes an individual's existence.

To ask what happens after we die is the same as wondering what existed before the universe began, or exists outside the universe now. There was no before; there is no outside; there is no after. It is all here and now.

In anything at all, perfection is finally attained not when there is no longer anything to add, but when there is no longer anything to take away.

What lies behind us and what lies before us are tiny matters compared to what lies within us.

For the infinitely little is equivalent to the infinitely great.

Objective reality is produced out of the collective memories of the human race while anomalous events are the manifestation of the individual will.

Efficiency requires that most of the time we ignore the rich texture of reality to attend to only one aspect of it: its use.

We live in a world in which there are no more links. We're just particles.

The challenge and the trick is to find wholeness among infinitude. One can live in the modern world not, as T.S. Eliot did, by grimly pitting his modern brokenness against a fantasy of perfection but by realizing that mastery and wholeness has always been a fantasy.

Time doesn't flow. The fact that we think time passes is just an accident of our nervous systems, of the way things look to us. In reality, time doesn't pass; we pass.

Time is what is needed to stop everything from happening at once.

Laughter, food, and sex focus on the here and now, such that past and future slip away.

The brain is like a radio, and consciousness is the broadcast signal: a wave.

Time and space exist on many levels; reality is plural and undetermined and confronts the possibility of multiple outcomes.

The world is a mirror of the mind's abundance.

If you travel into the past, you end up in the past of a different universe.

Every moment a beginning, every moment an end.

The weight of the soul--or life--is the energy difference between two closely related superpositions.

Consciousness requires language as well as emergence of a true self, along with concepts of the past and future.

The ability to have thoughts about thoughts allows an organism to go beyond perceived reality into the realm of the hypothetical.

Consciousness requires time passing; language requires time, and its control.

Insistence on complete logical clarification makes science impossible.

In the last few days of his life my father began to tell stories to people who came that were hallucinations, but he narrated them in a perfectly clear way. I realized that, at the end of your life, it doesn't matter whether something is objectively real to the outside world or not. There isn't any experential difference between what we call reality and hallucination.

The real reason that the universe is expanding is that the objects in it are staying in one place, and the space between them is growing.

You can make more money; you cannot make more time.

Lying is choosing a reality for yourself, and a universe for everyone else.

The world thus appears as a complicated tissue of events, in which connections of different kinds alternate or overlap or combine and thereby determine the texture of the whole.

Consciousness and identity are not a function of specific particles, because our own particles are constantly changing.

Now we must consider the perplexing matter cf time. How can it seem to move so quickly and yet so slowly too? How one summer to the next seems to pass while our heads were turned only briefly away, yet an eternity can elapse in an awkward moment though in reality not more than one minute has passed.

I believe in unlikely and unexpected affinities transacted between objects in a room. Or between people.

Each person's unique self is both imprinted in and manifest by the possessions he or she acquires over a lifetime.

Newton's theory employs action at a distance, a gravitational force that acts instantaneously. It anticipates quantum entanglement.

We lessen our fear of randomness by identifying regularities that do not exist. Belief dissociates from evidence.

The universe and life are based on chance, but not accident. Forces of chance and antichance coexist in a complementary relationship. The random element is entropy, the agent of chaos, which destroys meaning. The nonrandom element is information.

When people stop believing in something, they do not believe in nothing; they believe in everything.

If one believes consciousness is widely diffused in nature, it is hard to have much faith in science. The immediate causes of events, which science detects, offer only partial explanations to an observer who suspects some final purpose--or no purpose--is really shaping what happens.

Entangled particles are identical entities that share common origins and properties, and remain in instantaneous touch with each other, no matter how wide the physical gap between them.

It should make no difference whether the correlation between twin particles occurs when they are separated by a few meters or by the entire universe.

Einstein was wrong and action at a distance is real.

Decoherence must cause the universe to somehow split in two, spawning this world and another parallel "mirror world."

The past is not fixed but alters according to present decisions.

Procrastination keeps superpositions alive.

Strangeness lay in ordinary moments
Placed against a background
—Is there another word for "eternity"?—
Of the impersonal; curved space

Foaming with brief particles
As you leave the room

There may be multiverses, separate universes that cooled down differently, ending up governed by different laws and defined by different numbers.

The most radical aspect of quantum mechanics is that it tells us the world is essentially statistical.

Does chance or fate control our lives? The answer is as relative as time. If you're in your life, chance. Viewed from the outside, it's fate all the way.

Reality is a rounding error.

As soon as a system is reduced to its tiniest components, something is lost.

In small subatomic systems, uncertainty of position and momentum is significant. In larger, "real-world" objects, the quantitative amount of uncertainty becomes insignificant. Consequently, for human beings, causality and traditional determinism reign. Yet subatomic phenomena are scientifically significant in man, and to this extent causality governing him is weak, and he embodies both mechanical fate and statistical potentiality.

We can't ever know exactly who we are or why we do what we do; yes, we all exist in a state of contingency.

The principle of complementarity exists in our everyday lives: ways of looking at the things that happen to us and the people with whom we're involved that can never be reconciled.

There are as many realities as there are points of view, one for every person on earth.

Statistics can do nothing with a single piece of data. An isolated event has no meaning. Context is required.

The only thing that links you with your younger self is your memory. It's seated in your brain, the only organ that doesn't renew itself. While the rest of your body is in a constant state of reconstruction, you drag your mind along like an attic full of increasingly dusty memories. What you are is none other than a memory of yourself.

Reality in a material form requires the sense of time passing; otherwise it is merely a snapshot.

We do not experience the world as a superposition of possibilities, but only as a one-at-a-time sequence of definite actualities.

Quantum entanglement is "spooky action at a distance."

Time moves in relation to the amount of chaos. As order increases, time periods between salient events decrease. When there is a lot of chaos, it takes more time for significant events to occur.

Instability is inescapable; contain it, don't conceal it.

There is no such thing as importance or proportion. The little and the large matter the same.

Macroscopic objects do not appear as a smudge of possibilities because they have decohered by continuously jostling against the surrounding environment.

For large scale matter, time is an illusion involving laws of chance as applied to large numbers of atoms.

Consciousness observing a quantum uncertainty causes quantum decoherence.

The sound of the world collapsing and decohering is silence.

I can't take my body through time and space, but I can send my mind.

If the complex pattern of neurons firing in our brain has anything to do with consciousness and how we form our thoughts and perceptions, then decoherence of our neurons ensures that we never perceive quantum superpositions of mental states. In essence, our brains inextricably interweave the subject and the environment, forcing decoherence on us.

A way of being in the world while looking at it must require a quantum point of view; that is, all points of view at once.

Would it be fair to think of personality quarks, not quirks? Fundamental particles not just of nature in general, but of our own individual nature and peculiarities?

Particles can exist in high and low energy states simultaneously, like being happy and sad at once. Indeed, think of quantum matter as having multiple personalities that can all act out at once.

Once we admit that there is room for newness--that there are vastly more conceivable possibilities than realized outcomes--we must confront the fact that there is no special logic behind the world we inhabit.

Knowledge exists outside normal human consciousness--indeed, outside humans altogether--that is attainable if you disable the everyday faculties of reason and sensory perception.

Relationships cannot exist independent of time; thus, in any given instant the atom does not exist.

There is no "present" as such. Yet, paradoxically, we know the past only as a present memory and the future only as a present anticipation. There is, then, no real present and nothing but a real present.

There are many different realities coexistent all at once, and we can change and influence reality by our personal perception of it as well as by our behavior.

In 1901 light from a star 100 light-years away began its trip to your eyes. Did it know you would see it? Or would such light have emerged even if you'd never been born?

Quantum systems exhibit "non-local" behavior. Classical systems act only locally; that is, for a cause here to produce an effect there, an intervening mechanism must link the two.

Every life in some way affects every other life across space and time. Cause and effect are not linear, but instead embedded in our being. History is not linear.

A sharp distinction between animate and inanimate matter cannot be made. The stability of a living organism is a stability of process or function rather than form.

Time is the great mystery, more than any emotion, even love. It is the essential human mystery.

Words are the collapsed form of our quantum realities; to speak is to decohere.

Every portion of the holographic universe enfolds the whole. If we knew how to access it, we could find the Andromeda galaxy in the thumbnail of our left hand.

The failure of prediction permits us to act as if our choices make a difference.

98% of the atoms in the body are replaced annually. The stomach lining every week, the skin every month, the liver every six weeks. After five years, every atom is new. At age 30, our baby photo shows us six bodies ago. We are not abiding stuff, but patterns perpetuated.

Everything has two aspects: a normal one that we almost always see and which is seen by other people in general; the other, the spectral or metaphysical which can only be seen by rare individuals in moments of clairvoyance or metaphysical abstraction, just as certain bodies that exist within matter that cannot be penetrated by the sun's rays appear only under x-ray.

Extreme pleasure leaves one oblivious to the passage of time.

Beginners in the study of classical Greek are often troubled by the fact the the word <u>opiso</u> sometimes means "behind," sometimes "in the future." Speakers of English find this baffling because they are accustomed to thinking of themselves as moving through time. The Greeks, however, conceived of themselves as stationary, of time as coming up behind them, overtaking them, and then, still moving on, becoming the "past" that lay before their eyes.

It is not consciousness that is puzzling but matter. The existence of consciousness shows that we have a profoundly inadequate grasp of the nature of matter.

A religion for the 21st century: "The God of Small Things"

The world is a series of discrete, rapidly occurring, apparently seamless fields collapsing into a consensual (more or less) reality.

What happens to the life-force energy of living entities when they die? Since no energy can ever be lost, it must go somewhere. Is it possible such energy is converted into some invisible, undetectable form of matter?

Whenever you have more points of view, the chances of your getting the right solution are higher.

To understand life we must go beyond quantum theory, where physics and chemistry are necessary, to a place where history, perception, adaptation, and affection are equally essential.

Parallel universes may exist invisibly alongside ours, on their own membranes less than a millimeter away from ours. Such parallel universes could also be different sheets of our own universe folded back on itself.

Daylight leaks in and sluggishly I surface from my own dreams into the common dream and things assume again their proper places and their accustomed shapes.

Self and mind are the same thing.

The world of the subatomic meets with the world of the inconceivably vast in quantum mechanics, which the human mind can understand only mathematically.

Everything looks the same if you get close enough or far enough away; we live in-between.

Part of the Bayesian spirit is that you always have uncertainty about everything. The result of a Bayesian calculation is usually not a single number but a series of probabilities. At the core of Bayes' world-view is an admission that the world is rife with uncertainty.

It's always "now" for a photon.

The more you know, the less you need.

The more you understand, the less you need to know.

If both the live and the dead cat of Schrodinger exist simultaneously, the idea of linear time and reality collapses. Time and all realities must be layered, stacked infinitely high in any given instant, and our consiousness cuts a swath through them and calls it our life.

We can say that being awake or being conscious is nothing but a dreamlike state. It is a state that corresponds tightly to external reality, but it has no objective reality; as with a rainbow, you can perceive it but never actually touch or measure it.

To us it is clear that if a single moment could be seen complete it would disclose the whole.

A living organism is like a TV. DNA is the chipset and wiring. Behaviors are the programs. Just as programs don't arise inside the TV set, behaviors are a transmission, or morphic resonance, from previous similar organisms.

Telling the future by looking at the past assumes that conditions remain constant. This is like driving a car by looking in the rearview mirror.

The past is magnetic. It draws us in. We cannot help ourselves and, as with other things we cannot help in ourselves, we make up elaborate explanations, reasonable rational explanations, to chant away the powerful things that don't belong to us.

The energy of the vacuum remains one of the deep mysteries of science. We know from quantum mechanics that it is not empty.

We are all quantum fluctuations. That's the origin of all of us and of everything in the universe.

In a quantum world, not seeing something increases the probability of its being located elsewhere. Randomness must fail: order can be imposed on the world.

A titanic metamorphosis of a virtual particle into a real one occurred when a vacuum fluctuation 13 billion years ago converted a single virtual particle into a real infant universe. This conversion is known as the Big Bang.

All day long we give ourselves and others running accounts of why things are working the way they are. We look for order and reason always, even when they don't exist.

Space is not nothing. It is another name for the gravitational field of the universe. Space-time is a structural quality of the gravitational field. If you could turn off gravity with a switch, space-time would vanish.

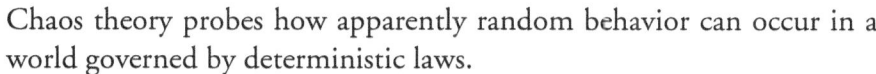

Chaos theory probes how apparently random behavior can occur in a world governed by deterministic laws.

Space and time have no objective reality. Yet the world is ultimately rational, and events within it must unfold according to the logic of cause and effect.

The chaos and the incongruities, it turns out, are part of the truth.

The idea of something existing without ever having begun is so alien to our framework of reality as to be virtually incomprehensible.

An atom is not a thing, but rather a set of forces operating in relationship to one another.

Our actual material content is changing constantly, and very quickly. We are rather like the patterns water makes in a stream. Thus, we should not associate our fundamental identity with specific sets of particles, but rather the pattern of matter and energy we represent.

The relation between what we see and what we know is never settled.

Space and time are modes by which we think, not conditions under which we live.

Doubt is the truth of our times.

Kant replaced reason with intuition. "If we remove the subject, space and time disappear; they cannot exist objectively but only in us."

Quantum mechanics is not itself a theory; rather, it is the framework into which all contemporary physical theory must fit. That framework requires the abandonment of determinism, classical physics, and absolute predictability.

I will know everything in the universe, how it was born, what it means, the secret of life and death- that there is no difference between them except the one we make in our minds.

Like many a young lover, quantum mechanics is deterministic in its head, but random in its heart.

A belief in prediction rests on a misguided faith in rock-hard causation and nature that is humanly rational. Randomness lies at the heart of the system. We have to live with some uncertainty.

Time was created with the world, it belongs to the world, therefore it did not exist before the universe existed. It is impossible to define in rational terms what is meant by the phrase "time has been created."

Our universe emerged from a point of infinite density of no physical size at Time Zero. This instant was blurred out by quantum mechanical effects, and dissolves away into a quantum fog. Before Time Zero, time did not exist.

When consciousness is derived entirely from the senses, you are entirely caught up in the here and now, trapped in the ultimate present.

Amid the seeming confusion of our mysterious world, individuals are so nicely adjusted to a system, and systems to one another, and to a whole, that, by stepping aside for a moment, a man exposes himself to a fearful risk of losing his place forever.

I wonder, maybe, if time stacks vertically, and there is no past, present, future, only simultaneous layers of reality. We experience our own reality at ground level. At a different level, time would be elsewhere. We would be elsewhere in time.

An insight or truth should be as valid in the distant future as it is in the present, if it is valid at all. Thus it can only be valued if considered with hindsight foreshadowed.

Superreality is a continually branching tree of possibilities in which everything that can happen actually does happen.

The map is not the territory; the word is not the thing; the belief about reality is not reality itself.

Video technology destroys our unconscious assumption of time's linearity. It unknots the rope of history. The teleology we read into simple chronological order is weakened by the endless playback and editing of recorded life. Chronological order no longer seems the invariably right way to read experience.

In the design of every subatomic particle, atom, molecule, cell, and thing, there appears to be something more at work than can be attributed to the parts alone.

This is not a world you escape, or one in which there is hope of redemption through coherence. It just is. The best you can do is understand its contingencies.

The universe is really more like a thought than a thing.

As to the mystery of how a force can occur between apparently disconnected bodies: the intervening space is somehow directly involved. A kind of tension exists in otherwise "empty" space that manifests itself by producing forces on objects in the vicinity.

One of the weird aspects of quantum mechanics is that something can simultaneously exist and not exist.

Life as manifested to us is a function of the asymmetry of the universe.

"Waves" and "particles" are things we can picture, images drawn from everyday experience, but that's all they are when we are talking about quantum mechanics. Light is not really made up of little particles, nor is it a wave; light is only completely described mathematically, an abstraction that corresponds to no physical picture at all.

One thing that all the processes of life seem to have in common is that they dissipate energy to create order.

Thought is one thing, reality another.

Natural language is superior to scientific language in adapting to changing knowledge, which is no surprise since the concepts of natural language are formed by the immediate connection with reality: they represent reality.

A theory of everything, from which both quantum theory and general relativity could be derived, would probably contain no concepts at all, but would be purely mathematical. Thus, anyone but an infinitely intelligent mathematician might as well give up ever trying to truly understand the world.

The ability to control a system without understanding everything about its parts or how or why it works characterizes our daily experience.

Miracles happen, not in opposition to nature, but in opposition to what we know of nature.

Is anyone there
if so
are you real
either way are you
one or several
if the latter
are you all at once

or do you
take turns…

The only thing that makes life possible is permanent, intolerable uncertainty: not knowing what comes next.

Life has no meaning, but we can give it a meaning.

Nobody really understands quantum mechanics.

NOTES

p.1   Only the incomprehensible: T. Parks, Destiny, 2000

One must always: W. Heisenberg, Physics and Philosophy, 1958, p.205

All human experience: N. Herbert, Quantum Reality, 1987, p.57

We cannot hope: J. Marks, Washington Times, April 9, 2000, p.B7

God created time: G. Wills, St. Augustine, 1999, p 91

p.2   Death is simply: J. Stirt, 2000

Look at it: J. Stirt 2000

The conscious mind: T. Sejnowski, Washington Post, March 20, 2000

Matter is thought: D. Mitchell, Ghostwritten, 1999, p.336

I could be: W. Shakespeare, Hamlet

Everything is not: J. Stirt, 2000

p.3   All things will: Anaxagoras, 460 B.C.

Nonsense and beauty: E.M. Forster, The Longest Journey, 1907

We live between: T. Parks, Destiny, 2000

Explanations are clumsy: I. Bergman

It is humanly: S. Freud in J. Lanchester, The Debt to Pleasure, 1996, p.223

There are many: N.Y. Times Book Review, 2000

p.4   We must decide: A. Carson, Poets & Writers, March/April 2001, p.28

Conscious life is: J. Stirt, 2000

All the different: G. Johnson, N.Y. Times, February 22, 2000

Take a living: J. Stirt, 2000

In biological terms: M. Dekkers, The Way of All Flesh, 1997, p.221

p.5   There is no: Schiller, The Death of Wallenstein, 1798

The future enters: R.M. Rilke, 1904

Time is downloaded: J. Winterson, Powerbook, 2000, p.121

Man is always: S. Budiansky, If A Lion Could Talk, 1998, p.58

Converts are not: J. Amato, Dust, p.174

p.6   We live in: J. Markoff, N.Y.Times, 2000

In the quantum: F. Close, Lucifer's Legacy, 2000, p.198

Consciousness is a: D. Bohm

The quantum self: M. Czikszentmihalyi, Wall St. Journal, 2000

p.7   Daily life is: J. Stirt, 2000

The world is: J. Rifkin, The Age of Access, p.193

Things and facts: W. Heisenberg, Physics and Philosophy, 1958, p.186

One reality does: J. Rosen, The Talmud and the Internet, 2000, p.138

What is it: I. Kalinowska, N. Y. Times, May 21, 2000

p.8   Quantum tunneling is: M. Browne, N.Y. Times, July 27, 1997

30% of: M. Tegmark, J. Wheeler, Scientific American, February 2001, pp.69&72

A barrier placed: R. Chiao, N.Y. Times, July 27, 1997

Humans can never: J. Stirt, 2000

Only atoms and: Democritus

p.9   To use quantum: J. Stirt, 2000

Magnetic resonance imaging: C. Platt, Wired Magazine, January 2000, p.207

If you can't: J. Stirt, 2000

Affect is the: J. Stirt, 2000

Traditional African thought: M. Jackson, Paths Toward a Clearing

p.10   Consciousness is a: D. Chalmers, in S. Greenfielc, The Private Life of the Brain, 2000, p.38

We tend to: T. Wolff, N.Y. Times, April 28, 2001

The ultimate impediment: L. MacFarquhar, New Yorker, May 29, 2000, p.105

Quantum thinking eliminates: J. Stirt, 2001

Coincidence is a: J. Stirt, 2000

p.11   Contagious magic is: J. Frazier

Quantum reality consists: N. Herbert, Quantum Reality, 1987, p. 248

Quantum physics is: M. Browne, N.Y Times, July 27, 1997

To change the: J. Stirt, 2000

The more intense: S. Greenfield, The Private Life of the Brain, 2000

p.12   Chaos is the: J. Stirt, 2000

Life is but: S.T. Coleridge

I do not: V. Nabokov

Consciousness is a: J. Stirt, 2000

Quantum mechanics adopted: R. Kurzweil, The Age of Spiritual Machines, p.62

p.13   Our judgment does: S. Budiansky, If A Lion Could Talk, 1998, p.58

An entity: R. Penrose in R. Kurzweil, The Age of Spiritual Machines, p.118

Matter and consciousness: J. Stirt, 2000

The atom is: F. Close, Lucifer's Legacy, 2000, p.125

Old science sought: J. Rifkin, The Age of Access, p. 193

p.14   Interaction between individuals: J. Stirt, 2000

In quantum reality: J. Winterson, Powerbook, 2000, p.63

Scripts, sets, personae: J. Stirt, 2000

Quantum man must: J. Stirt, 2000

There is no: P. Romer, in M. Lewis, The New New Thing, p. 252

p.15   The law of: W. Heisenberg, Physics and Philosophy, 1958, p.88

Objects carry with: J. Stirt, 2000

The power of: A. Hartman, Net Ready, p. 150

If each neuron: J. Stirt, 2000

p.16   Emotion, the most: S. Greenfield, The Private Life of the Brain, 2000

What might be: W. Mosley, N.Y. Times, July 3, 2000

We lead our: J. Podeswa, N.Y. Times, July 9, 2000

Man has a: S. Budiansky, If A Lion Could Talk, 1998, pp.xvii-xviii

Superpositions have a: Economist, July 8, 2000, p. 83

p.17   What would quantum: J. Stirt, 2000

How much of: J. Stirt, 2000

It's only because: J. Saramago, All the Names, 1999, pp.36-7

In the unpredictability: J. Stirt, 2000

The ultimate quantum: J. Stirt, 2000

p.18   The quantum reptile: J. Stirt, 2000

As long as: J. Stirt, 2000

In the realm: K. Chang, N.Y. Times, July 11, 2000

The probabilistic nature: M. Browne, N.Y. Times, July 27, 1997

p.19   Everyday life is: K. Chang, N.Y. Times, July 11, 2000

The illusion of: D. Weiner, Battling the Inner Dummy, pp. 318-19

We cannot see: S. Odenwald, Washington Post, May 14, 1997

Life is a: J. Stirt, 2000

To explain the: J. Campbell, Grammatical Man, 1982, p.16

p.20   The principle of: H. Pagels, The Cosmic Code

Quantum intuition is: J. Stirt, 2000

Quantum reality can: J. Stirt, 2000

Thought can exceed: J. Stirt, 2000

Life is a: J. Stirt, 2000

All work and: I. Murdoch

p.21   One must have: E. Ionesco

Everything is true: R.L. Stevenson

When you can't: J. Stirt, 2000

Intellectuals have a: N.Y. Times

p.22   The body is: E. Becker

When someone invests: M. Csikszentmihalyi, The Meaning of Things, p.8

Objects are more: T. Folger, Discovery Magazine, December 2000

Many entities that: N. Herbert, Quantum Reality, 1987, p.171

p.23   Death has a: N.Y. Times

You were dead: N. French, Beneath the Skin, 2000, p.314

Dividing the universe: D. Bohm

Time present and: T.S. Eliot, Four Quartets, 1935

Human consciousness collapses: D. Mitchell, Ghostwritten, 1999, p.348

p.24   The flow of: F. Close, Lucifer's Legacy, 2000, p.216

Entropy happens in: J. Stirt, 2000

The transition from: W. Heisenberg, Physics and Philosophy, 1958, p.54

Nothing is nothing: Aboriginal epigram

The truest and: R. Eder, N.Y. Times, April 11, 2001

p.25   What exists and: J. Winterson, Powerbook, 2000, p.129

Religion, morality, and: J. Amato, Dust

Fundamentally, is the: L. Smolin, N.Y. Times

"Nothing" is the: E. Dickinson

Once the: S. Holmes in J. Lanchester, The Debt to Pleasure, 1996, p. 181

p.26   That matter is: L. MacFarquhar, New Yorker, May 29, 2000, p.105

The "primordial soup": J. Stirt, 2000

Certain moments have: MacIver, N.Y. Times, 2000

My wish is: J. Stirt, 2000

Physical objects of: M. Darrieuscecq, My Phantom Husband

p.27   Apparently empty space: J. Glanz, N.Y.Times, February 9, 2000, p.18

Time is the: N. Kozyrev, 1967

What is it: S. Budiansky, If A Lion Could Talk, 1998, p.192

Man has invented: I. Bergman

That which constitutes: G. De Chirico, 1919

p.28   The universe itself: G. Johnson, N.Y.Times, February 22, 2000

The world can: P. Davies

The past and: D. Deutsch, The Fabric of Reality, 1997, p.288

Empty space is: K. Chang, N.Y.Times, February 9, 2001, p.19

At the root: F. Dyson

p.29   Religion occurs when: S. Greenfield, The Private Life Of The Brain, 2000, p.75

Magic, religion, and: J.R. Frazier, The Golden Bough, 1915

Time is a: G. Wills, St. Augustine, 1999, p.90

Time exists only: J. Stirt, 2000

p.30   There is no: T. Folger, Discovery Magazine, December 2000

Self-awareness is a: V. Nabokov, Lolita

A human experiences: A. Einstein, in N. Herbert, Quantum Reality, 1987, p.250

Without eyes you: M. Dekkers, The Way of All Flesh, 1997, p.221

Since at the: T. Ferris, N.Y. Times Magazine, September 29, 1996

p.31   The apparently accidental: N.Y.Times

The only truths: J. Stirt, 2000

In collapsing space: J. Stirt, 2000

Since mass and: W. Heisenberg, Physics and Philosophy, 1958

p.32   Space has always: S. Odenwald, Washington Post, May 14, 1997

The Corinthians asked: N.Y. Times

You attribute this: J.P. Sartre

The difference between: J. Baudrillard in J. Rifkin, The Age of Access, p.197

p.33   In saving the: J. Holt, Wall St. Journal, January 2ॐ, 1997

The Western objective: R. Kurzweil, The Age of Spiritual Machines, 2000, p.82

Man might not: S. Redmond, N.Y.Times, 1980

Nothing completely disappears: Christian Science Monitor, 1976

p.34   Consciousness may arise: R. Penrose, Time Magazine, July 17, 1995

The universe as: A. Penzias

New and old: A. Bitov

The painter's task: P. Cezanne

I do not: G. Balanchine

Magic/superstition-control: J. Stirt, 2000

p.35   Time does not: S. Saunders, N.Y.Times, March 23, 2000

Language must make: J. Stirt, 2000

Fear of the: R.M. Rilke, 1904

To ask what: J. Stirt, 2000

p.36   In anything at: A. St.-Exupery

What lies behind: R.W. Emerson

For the infinitely: M. Maeterlinck, The Life of an Ant

Objective reality: J.P. Briggs & F.D. Peat, The Looking Glass Universe, 1984, p.87

Efficiency requires that: Y.-F. Tuan, Morality and Imagination

p.37   We live in: M. Houellebecq

The challenge and: J. Rosen, The Talmud and the Internet, 2000, pp.35-6

Time doesn't flow: M. Crichton, Timeline, 1999, p.108

Time is what: F. Close, Lucifer's Legacy, 2000, p.229

Laughter, food, and: S. Greenfield, The Private Life of The Brain, 2000, p. 154

p.38   The brain is: J. Stirt, 2000

Time and space: Economist, October 2000

The world is: J. Winterson, Powerbook, 2000, p.263

If you travel: D. Deutsch, The Fabric of Reality, 1997, p.319

Every moment a: M. Salzman, Lying Awake, 2000

The weight of: Scientific American, October 2000

p.39   Consciousness requires: G. Edelmar, A Universe of Consciousness, 2000

The ability to: S. Budiansky, If A Lion Could Talk, 1998, p.163

Consciousness requires time: J. Stirt 2000

Insistence on complete: W. Heisenberg, Physics and Philosophy, 1958, p.86

In the last: K. Chalfant

p.40   The real reason: S. Odenwald, Washington Post, May 14, 1997

You can make: J. Stirt, 2000

Lying is choosing: J. Stirt, 2000

The world thus: W. Heisenberg, Physics and Philosophy, 1958

Consciousness and identity: R. Kurzweil, The Age of Spiritual Machines, p.54

p.41   Now we must: D. Di Blasi, Prayers of an Accidental Nature, 1998, p.159

I believe in: D. Antrim, The Verificationist, 2000, p.80

Each person's unique: Hegel, in J. Rifkin, The Age of Access, p.215

Newton's theory employs: M. Gell-Mann, The Quark and the Jaguar, 1994, p.87

We lessen our: M. Gell-Mann, The Quark and the Jaguar, 1994, p.283

p.42   The universe and: J. Campbell, Grammatical Man, 1982, p.28

When people stop: F. Fernandez-Armesto, Truth, 1997, p.3

If one believes: F. Fernandez-Armesto, Truth, 1997, pp.144-45

Entangled particles are: M. Browne, N.Y. Times, July 22, 1997

p.43   It should make: N. Gisin, N.Y. Times, July 22, 1997

Einstein was wrong: M. Browne, N.Y. Times, July 22, 1997

Decoherence must cause: G. Johnson, N.Y. Times, February 22, 2000

The past is: N. Herbert, Quantum Reality, 1987, p.166

Procrastination keeps superpositions: J. Stirt, 2000

Strangeness lay in: J. Koethe, The New Yorker, May 8, 2000

p.44   There may be: M. Rees, Just Six Numbers

The most radical: J. Marsh, Washington Times, April 9, 2000, p.B7

Does chance or: D. Mitchell, Ghostwritten, 1999, p.283

Reality is a: J. Stirt, 2000

p.45   As soon as: S. Greenfield, The Private Life of the Brain, p.9

In small subatomic: W. Heisenberg, Physics and Philosophy, 1958, p.18

We can't ever: N.Y.Times, April, 2000

The principle of: N.Y.Times, April, 2000

There are as: J. Ortega y Gasset

p.46   Statistics can do: J. Campbell, Grammatical Man, 1982, p.28

The only thing: M. Dekkers, The Way of All Flesh, 1997, p.221

Reality in a: J. Stirt, 2000

We do not: N. Herbert, Quantum Reality, 1987, p.243

Quantum entanglement is: A. Einstein

p.47   Time moves in: R. Kurzweil, The Age of Spiritual Machines, p.29

Instability is inescapable: R. Koolhas

There is no: J. Stirt, 2000

Macroscopic objects do: K. Chang, N.Y.Times, July 11, 2000

For large scale: F. Close, Lucifer's Legacy, 2000, p.229

p.48   Consciousness observing a: R. Kurzweil, The Age of Spiritual Machines, p.118

The sound of: J. Stirt, 2000

I can't take: J. Winterson, Powerbook, 2000, p.63

If the complex: M. Tegmark, J. Wheeler, Scientific American, February 2001, p.7

A way of: J. Stirt, 2000

p.49   Would it be: J. Stirt, 2000

Particles can exist: J. Yaukey, USA Today, July 12, 2000

Once we admit: P. Romer

Knowledge exists outside: L. MacFarquhar, New Yorker, May 29, 2000, p.111

p.50   Relationships cannot exist: J. Stirt, 2000

There is no: G. Wills, St. Augustine, 1999, p.91

There are many: P.K. Dick

In 1901 light: J. Stirt, 2000

Quantum systems exhibit: T. Ferris, N.Y.Times Magazine, September 29, 1996 p.51   Every life in: L. Lescaze, Wall St. Journal, 1993

A sharp distinction: W. Heisenberg, Physics and Philosophy, 1958, p.154

Time is the: T. Mallon, N.Y.Times, January 30, 1997

Words are the: J. Stirt, 2000

Every portion of: M. Talbot, The Holographic Universe, 1992, p.50

p.52   The failure of: A. Schlesinger, Wall St. Journal

98% of the: B. Lemley, Washington Post, February 23, 1986

Everything has two: G. De Chirico, 1919

Extreme pleasure: S. Greenfield, The Private Life of the Brain, 2000, p.102

p.53   Beginners in the: C. Kluckhohn, Mirror For Man

It is not: G. Strawson, N.Y.Times, July 11, 1999

A religion for: A. Roy

The world is: J. Stirt, 2000

What happens to: A. Zweibel, N.Y. Times

p.54   Whenever you have: A. Greenspan

To understand life: W. Heisenberg, Physics and Philosophy, 1958, p.104

Parallel universes may: Scientific American, August 2000, p.67

Daylight leaks in: J.L. Borges

p.55   Self and mind: S. Greenfield, The Private Life of the Brain, 2000

The world of: J. Amato, Dust, 2000, p. xi

Everything looks the: J. Stirt, 2000

Part of the: M.I. Jordan, N.Y.Times, April 28, 2001

It's always now: J. Stirt, 2000

p.56   The more you: Aboriginal saying

The more you: J. Stirt, 2000

If both the: J. Stirt, 2000

We can say: R. Llinas, Time Magazine, July 17, 1995

To us it: W.S. Merwin

p.57   A living organism: R. Sheldrake & A. Postman

Telling the future: H. Brody, Wired Magazine

The past is: J. Winterson, Powerbook, 2000, p.197

The energy of: M. Turner, N.Y.Times, January 21, 1997

p.58   We are all: J. Bahcall, N.Y. Times, January 21, 1997

In a quantum: J. Stirt, 2000

A titanic metamorphosis: M. Browne, N.Y.Times, January 21, 1997

All day long: M. Gazzaniga, The Mind's Past, 1998, p.157

Space is not: S. Odenwald, Washington Post, May 14, 1997

p.59   Chaos theory probes: N.Y. Times, October, 2000

Space and time: A. Einstein

The chaos and: J. Rosen, The Talmud and the Internet, 2000

The idea of: A. Zweibel, N.Y. Times

An atom is: J. Rifkin, The Age of Access, p.192

p.60   Our actual material: R. Kurzweil, The Age of Spiritual Machines, pp.54-5

The relation between: J. Berger

Space and time: A. Einstein

Doubt is the: F. Fernandez-Armesto, Truth, 1997, p.206

Kant replaced reason: F. Fernandez-Armesto, Truth, 1997, p.169

p.61   Quantum mechanics is: M. Gell-Mann, The Quark and the Jaguar, 1994, p.6

I will know: T. Moran, The World I Made For Her, 1998

Like many a: P. Halpern, The Pursuit of Destiny, 2001

A belief in: D. Teresi, Wall St. Journal, January 11, 2001

Time was created: W. Heisenberg, Physics and Philosophy, 1958, p.125

p.62   Our universe emerged: S. Odenwald, Washington Post, May 14, 1997

When consciousness is: S. Greenfield, The Private Life of the Brain, 2000,
        p.101

Amid the seeming: N. Hawthorne

I wonder, maybe: J. Winterson, Powerbook, 2000, pp.219-20

p.63   An insight or: J. Stirt, 2000

Superreality is a: N. Herbert, Quantum Reality, 1987, p.173

The map is: A. Korzybski, Science and Sanity

Video technology destroys: A. Marlowe, How to Stop Time, 1999, p.107

p.64   In the design: B. Hubbard, Conscious Evolution, 1998, p.91

This is not: T. Luhrmann, N.Y.Times Book Review, January 21, 2001, p.18

The universe is: J. Wheeler, in B. Hubbard, Conscious Evolution, p.43

As to the: F. Close, Lucifer's Legacy, 2000, p. 138

One of the: M. Browne, N.Y. Times, July 22,1997

p.65   Life as manifested: L. Pasteur, 1860

        "Waves" and "particles": S. Budiansky, If A Lion Cou d Talk, 1998, pp.73-4

        One thing that: Economist, January 27, 2001, p.79

        Thought is one: C. McGinn, The Mysterious Flame, 1999, p.65

        Natural language is: W. Heisenberg, Physics and Philosophy, 1958, p.200

p.66   A theory of: M. Tegmark, J. Wheeler, Scientific American, February 2001, p.75

        The ability to: J. Stirt, 2001

        Miracles happen, not: St. Augustine

        Is anyone there: To The Soul, W.S. Merwin, New Yorker, February 5, 2001, p.40

p.67   The only thing: U. LeGuin, The Left Hand of Darkness, 1969, p.72

        Life has no: J. Stirt, 2001

        Nobody really understands: R. Feynman.

0-595-21159-3